A LOOK AT EARTH SCIENCE

ASTRONOMY

BY LUKE HARASYMIW

Gareth Stevens PUBLISHING

Please visit our website, www.garethstevens.com. For a free color catalog of all our high-quality books, call toll free 1-800-542-2595 or fax 1-877-542-2596.

Library of Congress Cataloging-in-Publication Data

Names: Harasymiw, Luke, author.
Title: Astronomy / Luke Harasymiw.
Description: New York : Gareth Stevens Publishing, [2025] | Series: A look at Earth science | Includes bibliographical references and index.
Identifiers: LCCN 2024006596 (print) | LCCN 2024006597 (ebook) | ISBN 9781482467123 (library binding) | ISBN 9781482467116 (paperback) | ISBN 9781482467130 (ebook)
Subjects: LCSH: Astronomy–Juvenile literature.
Classification: LCC QB46 .H2187 2025 (print) | LCC QB46 (ebook) | DDC 520–dc23/eng/20240314
LC record available at https://lccn.loc.gov/2024006596
LC ebook record available at https://lccn.loc.gov/2024006597

First Edition

Published in 2025 by
Gareth Stevens Publishing
2544 Clinton Street
Buffalo, NY 14224

Designer: Jennifer Schoembs
Editor: Therese M. Shea

Photo credits: Cover, back cover, left pages, series art (universe) pp. 1, 29 Vadim Sadovski/Shutterstock.com; p. 5 New Africa/Shutterstock.com; p. 7 (inset) File:Su Song Star Map 2.JPG/Wikimedia Commons; p. 7 (main) File:Nebra disc 1.jpg/Wikimedia Commons; p. 9 sripfoto/Shutterstock.com; p. 11 alexaldo/Shutterstock.com; p. 13 File:Comet C2020F3 NEOWISE over California desert landscape.png/Wikimedia Commons; p. 15 Mopic/Shutterstock.com; p. 17 Ingo Bartussek/Shutterstock.com; p. 19 File:Galileo Galilei (1564-1642) RMG BHC2700.tiff/Wikimedia Commons; p. 21 (inset) File:JKepler.jpg/Wikimedia Commons; p. 21 (main) File:Nikolaus Kopernikus.jpg/Wikimedia Commons; p. 23 (top) AstroStar/Shutterstock.com; p. 23 (bottom) Pavel Tvrdy/Shutterstock.com; p. 25 KENNY TONG/Shutterstock.com; p. 27 (main) NASA, ESA, CSA, STScl; p. 27 (inset) File:James Webb Space Telescope 2009 top.jpg/Wikimedia Commons.

Printed in the United States of America

CONTENTS

Words in the glossary appear in **bold** type the first time they are used in the text.

LOOK UP!

While the sun is up, it's easy to forget what's beyond Earth's **atmosphere**. But if you're outside on a clear night, the stars and **planets** in the night sky get your attention. Astronomy is the study of everything outside Earth's atmosphere.

MAKE THE GRADE

Scientists consider astronomy to be a part of the study of Earth science. Astronomy helps us understand more about our planet, such as how the moon and sun cause tides.

WHAT WE CAN SEE

A scientist who studies astronomy is called an astronomer. The earliest astronomers didn't have the tools that today's astronomers have. They could only study what they could see with their own eyes. This is called observational astronomy. They recorded objects, movements, and **patterns** in the night sky.

star map created in China around 1092 CE

MAKE THE GRADE

People have been recording the night sky for more than 3,000 years. The Nebra Sky Disk, shown here, dates back to 1600 BCE.

OUR SOLAR SYSTEM

Earth is one of eight planets in our planetary system. A planetary system is a star and the planets that orbit, or move around, it. We call our planetary system the solar system. It's one of more than 3,000 planetary systems in the Milky Way **galaxy**.

MAKE THE GRADE

Astronomers have learned the Milky Way galaxy is one of billions of galaxies in the universe. The universe is all of space and everything in it.

GRAVITY

Gravity is the force that keeps the solar system in motion. The sun has about 99 percent of all the **mass** in the solar system. That mass keeps the planets in orbit. But planets are large enough to have their own gravity too.

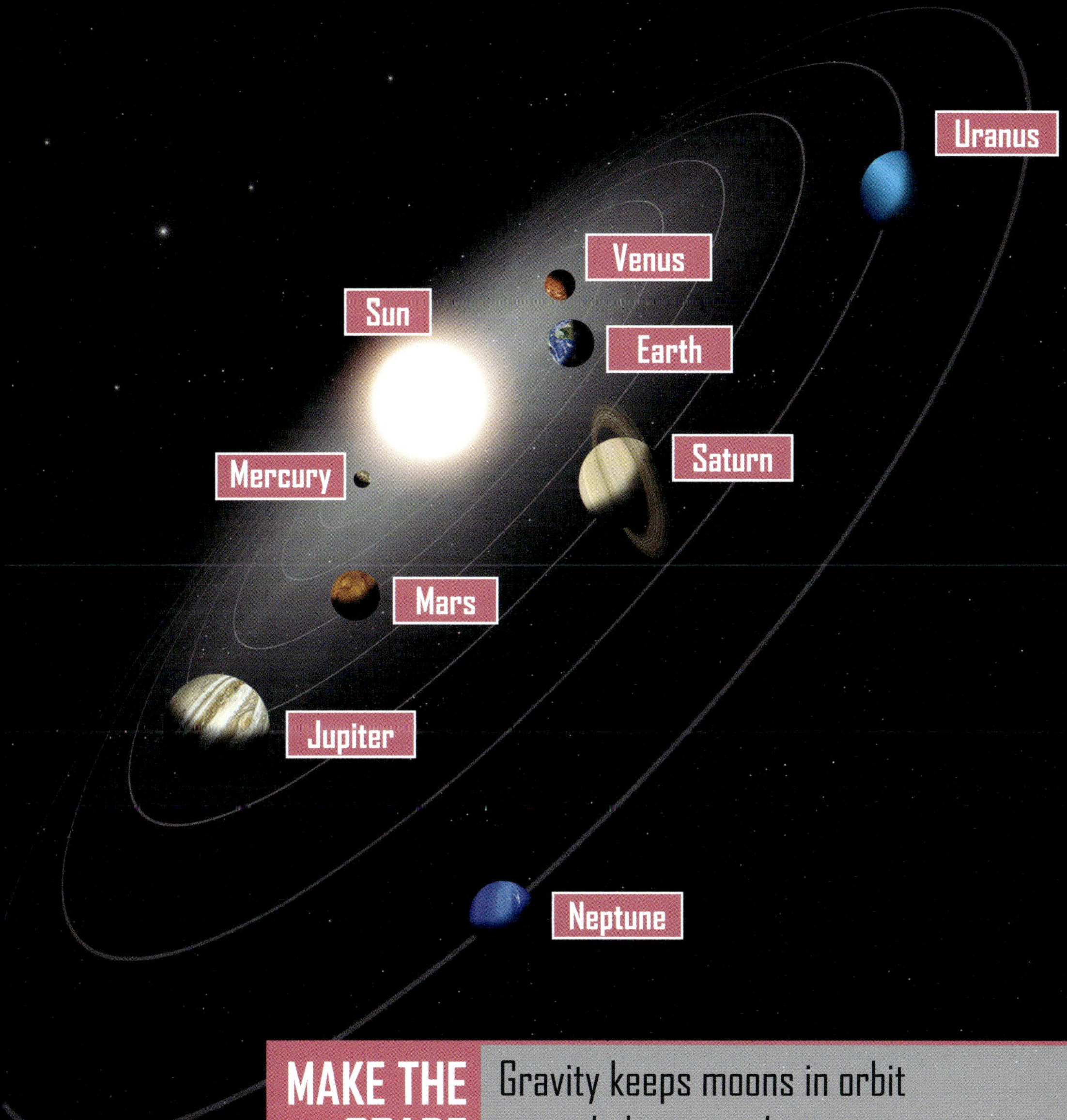

MAKE THE GRADE

Gravity keeps moons in orbit around planets and gives planets their round shape.

COMETS

A comet is a piece of ice, rock, dust, and gases that orbits the sun. Comets have an oval orbit. As they get closer to the sun, their gases heat up and form a glowing cloud called a coma. Gas and dust form a long tail behind the comet.

MAKE THE GRADE

A comet's tail can be as long as the distance from Earth to the sun, or 93 million miles (150 million km) long!

ASTEROIDS

Asteroids are large chunks of rock and metal that orbit the sun. Some are formed by collisions, or crashes, in space. Others are left over from a planet forming. Most large asteroids are between 62 miles (100 km) wide and 620 miles (1,000 km) wide.

MAKE THE GRADE

Most asteroids in our solar system are found between Mars and Jupiter. This area is called the asteroid belt.

METEOROIDS

Smaller pieces of rock in space are called meteoroids. If they enter Earth's atmosphere, they're called meteors. They usually burn up before they hit the ground. They look like a streak of light. Sometimes they're called falling stars. When meteors hit the ground, they're called meteorites.

MAKE THE GRADE

Astronomers study what meteorites are made of to find out where they came from. Many meteorites seem to come from a space object colliding with the moon or with Mars.

FAMOUS ASTRONOMERS

Astronomers are the reason we have so much **information** about our solar system. Galileo Galilei (1564–1642) was a famous early astronomer. He made a telescope in 1609. Using this, he saw craters and mountains on the moon. He also observed four moons orbiting Jupiter.

MAKE THE GRADE

NASA sent a space **probe** called *Galileo* to Jupiter in 1989. It arrived in 1995. It sent back information about asteroids, Jupiter's moons, and Jupiter itself.

Nicolaus Copernicus (1473–1543) is called the founder of modern astronomy. In his writings, he stated that Earth spins around its axis, or center. He also said that Earth orbits the sun. At the time, many people believed the sun orbited Earth.

MAKE THE GRADE

Johannes Kepler (1571–1630) later proved Copernicus (above) was right about Earth orbiting the sun. Kepler also proved the planets have an oval-shaped orbit.

Johannes Kepler

TELESCOPES

Today's astronomers can study objects inside and outside our solar system. They do this with powerful **technology**. Very large telescopes use **mirrors** to gather and collect light from faraway objects in space. These telescopes are called **reflecting** telescopes. Some powerful ones are in large buildings called observatories.

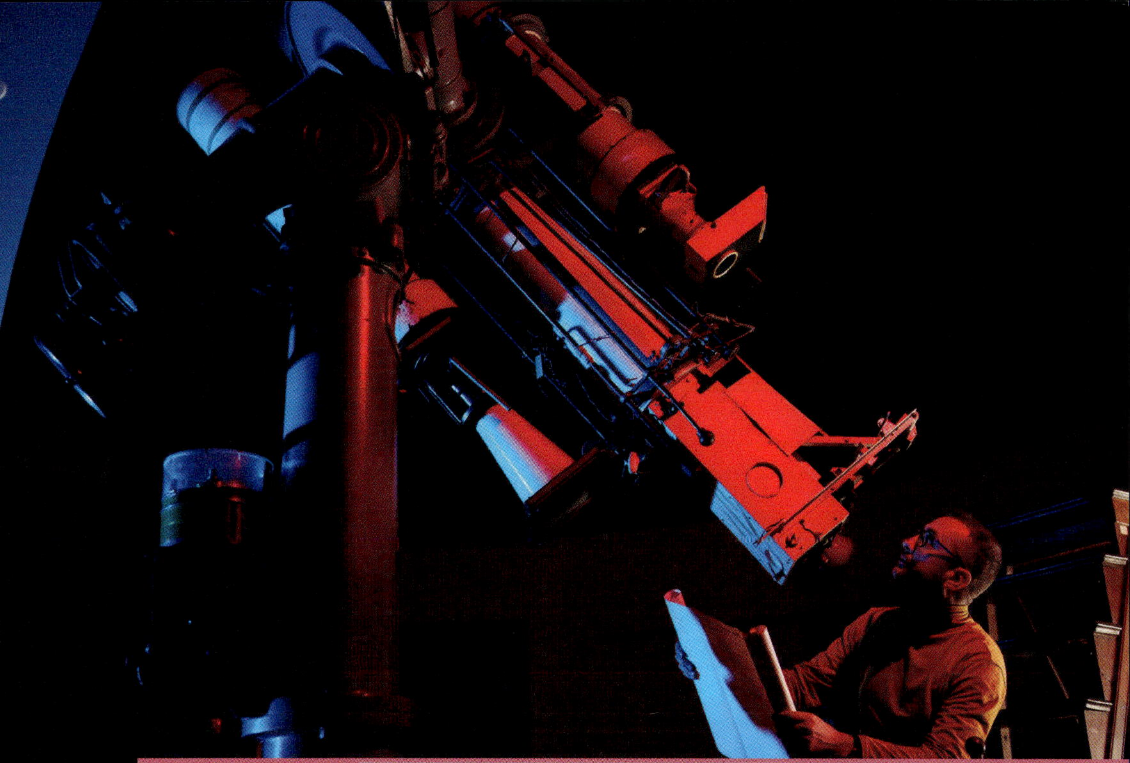

MAKE THE GRADE

Many smaller telescopes are refracting telescopes. They use pieces of glass called lenses to refract, or bend, light to make objects look closer. They aren't as powerful as reflecting telescopes.

observatory on Mauna Kea in Hawaii

Radio telescopes are large dishes with antennae. They collect waves of **energy**, called radio waves, from space. Astronomers use radio waves to find out what an object is made of, what it looks like, and if it's moving. Some radio telescopes detect, or find, radio waves from other galaxies.

MAKE THE GRADE

To make better images, or pictures, astronomers aim a number of radio telescopes at the same spot. This is called an array.

OUT OF THIS WORLD

Earth's atmosphere can make it hard for telescopes to get a clear view of space objects. Astronomers found a way around this. They put telescopes into space. The James Webb Space Telescope is the largest, most powerful space telescope ever built. It orbits the sun.

James Webb Space Telescope

MAKE THE GRADE

The James Webb Space Telescope (JWST) uses special cameras that find and create images using heat energy it detects. This JWST image shows the Carina Nebula, a cloud of gas and dust in our galaxy.

RECENT DISCOVERIES

Astronomers make new discoveries every year. In 2023, they found out Mercury is **shrinking**. They discovered Mars's spin is getting faster. They found 16 new moons around Saturn, bringing its moon count to 146. And those are just the discoveries in our solar system!

MAKE THE GRADE

As of 2023, astronomers have found more than 5,500 exoplanets, or planets outside our solar system.

MORE ASTRONOMY TOOLS AND TECHNOLOGY

spectrographs—*break light from an object into different colors, revealing the kinds of matter in it*

star charts—*map stars and planets, helping astronomers and others find space objects*

advanced camera systems—*take photos of light that can be seen with the human eye as well as light energy that can't be seen*

space probes—*spacecraft without a person in them sent out to gather facts about space and the objects in it*

interferometers—*use light waves and radio waves that can measure the size of space objects*

computers and computer programs—*can be used to model certain events in space, such as stars forming*

GLOSSARY

atmosphere: The mixture of gases that surround a planet.

energy: A power source.

galaxy: A large group of stars, planets, gases, and dust that form a unit within the universe.

gravity: The force that pulls objects toward the center of a planet or star.

information: Knowledge obtained from study or observation.

mass: The amount of matter in an object.

mirror: A piece of glass that shows the image in front of it.

NASA: Stands for the National Aeronautics and Space Administration. It's the part of the U.S. government in charge of the science and technology involving airplanes and space.

pattern: The way something happens over and over again.

planet: A large, round object in space that travels around a star.

probe: A machine used to collect information about space and send it back to Earth.

FOR MORE INFORMATION

Books

Betts, Bruce. *Astronomy for Kids: How to Explore Outer Space with Binoculars, a Telescope, or Just Your Eyes!* Emeryville, CA: Rockridge Press, 2021.

Poppele, Jonathan. *Stargazing for Kids: An Introduction to Astronomy.* Cambridge, MN: Adventure Publications, 2022.

Website

What Is Astronomy?
www.amnh.org/explore/ology/astronomy
Check out a lot of astronomy links, including a kid's guide to astronomy.

INDEX